PHYSICAL SCIENCE

CHANGES IN MATTER

By Christina Earley

A Stingray Book

SEAHORSE PUBLISHING

Teaching Tips for Caregivers and Teachers:

This Hi-Lo book features high-interest subject matter that will appeal to all readers in intermediate and middle school grades. It may be enjoyed by students reading at or above grade level as well as by those who are looking for age-appropriate themes matched with a less challenging reading level. Hi-Lo books are ideal for ELL readers, too.

Each book appeals to a striving reader's age and maturity level. Opportunities are provided for students to read words they already know while encountering a limited number of new, high-interest vocabulary words. With these supports in place, students will read more fluently while increasing reading comprehension. Use the following suggestions to help students grow as readers.

- Encourage the student to read independently at home.
- Encourage the student to practice reading aloud.
- Encourage activities that require reading.
- Establish a regular reading time.
- Have the student write questions about what they read.

Teaching Tips for Teachers:

Before Reading

- Ask, "What do I know about this topic?"
- Ask, "What do I want to learn about this topic?"

During Reading

- Ask, "What is the author trying to teach me?"
- Ask, "How is this like something I already know?"

After Reading

- Discuss how the text features (headings, index, etc.) help with understanding the topic.
- Ask, "What interesting or fun fact did you learn?"

TABLE OF CONTENTS

WHAT IS MATTER?

Matter is anything that has mass and takes up space.

It is made of **molecules** that move around.

Changes in matter can be **physical** or **chemical**.

STATES OF MATTER

There are three states of matter.

Solids have a definite shape.

Liquids take the shape of the container.

Gases can spread throughout any space.

FUN FACTS

Molecules in a solid are packed tightly and organized. Molecules in a liquid are close together. Molecules in a gas are spread far apart.

PHYSICAL CHANGE

A physical change is a change in the physical properties of an object.

The matter is the same material after the change.

The change can be reversed.

Cutting grass and crumpling paper are physical changes.

FUN FACTS

Erosion creates a physical change in the land.

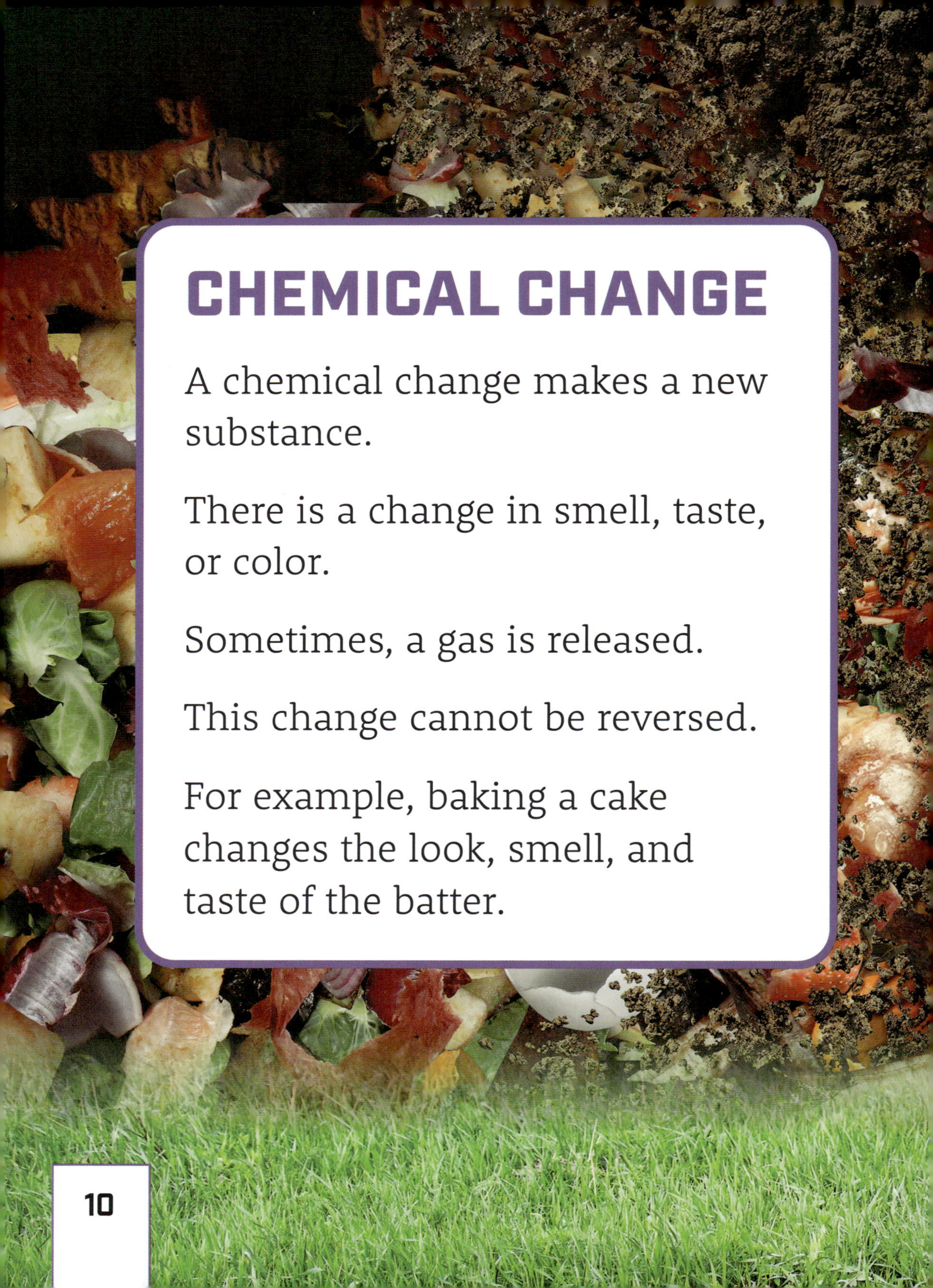

CHEMICAL CHANGE

A chemical change makes a new substance.

There is a change in smell, taste, or color.

Sometimes, a gas is released.

This change cannot be reversed.

For example, baking a cake changes the look, smell, and taste of the batter.

FUN FACTS

Plants *grow*, produce fruit, and become compost for new plants as a result of chemical changes.

TEMPERATURE CHANGE

Temperature can change the state of matter.

The melting point is the temperature when a solid changes to a liquid.

The freezing point is the temperature at which a liquid changes to a solid.

The boiling point is the temperature when a liquid changes to a gas.

LAW OF CONSERVATION OF MATTER

The Law of Conservation of Matter states that the amount of matter stays the same even when it changes form.

Matter cannot appear or disappear.

It only changes from one form to another.

Things cannot be created magically or **destroyed**.

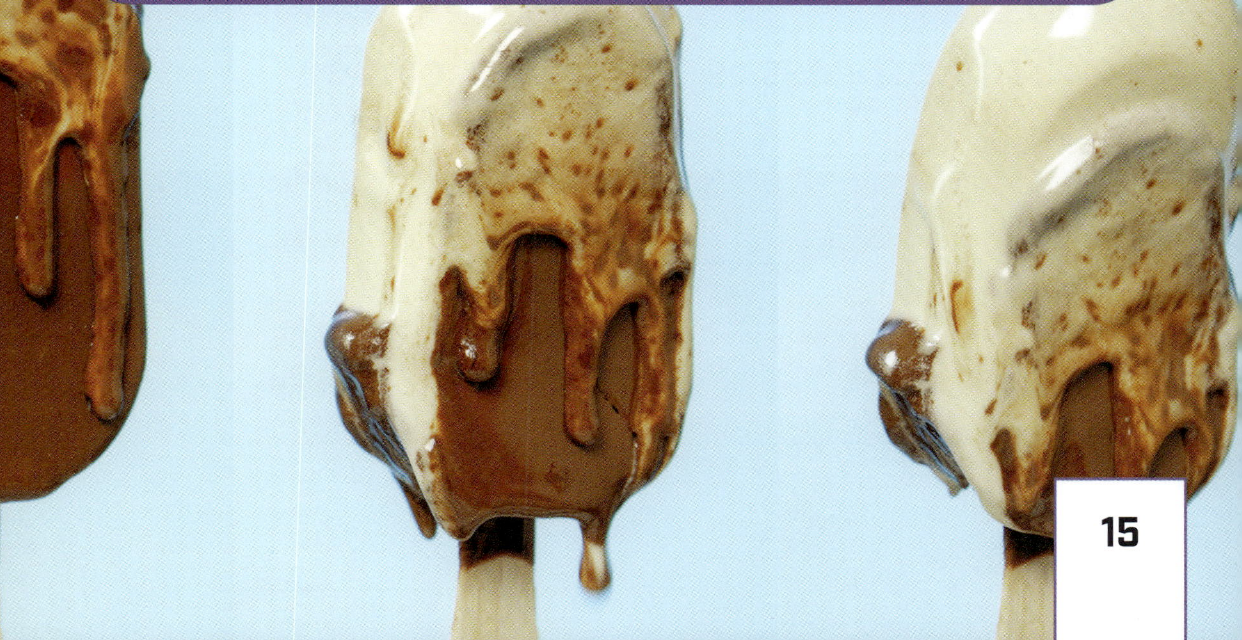

THE WATER CYCLE

Energy from the sun causes liquid water in bodies of water to **evaporate**, or turn into a gas.

As the gas rises into the sky, condensation cools it back into liquid drops that make up clouds.

Precipitation happens when enough water has condensed.

Rain, snow, and hail fall from clouds and are collected in bodies of water.

This process is the water **cycle**.

FUN FACTS

We could be drinking the same water that dinosaurs drank.

CAREER: CHOCOLATIER

Chocolatiers make and sell candies made from chocolate.

They get the cocoa beans ready to make the chocolate.

To make candies, chocolate needs to be tempered, or heated and cooled.

Some even create chocolate sculptures.

INVESTIGATE: WATER CYCLE BOTTLE

Materials:

- Empty plastic water bottle with cap
- Plastic cup
- Permanent marker
- Ice cubes
- Blue food coloring
- Water

Procedure:

(1) Remove any labels from the water bottle.

(2) Hold the bottle with the cap facing the floor.

(3) Use the marker to draw clouds at the new top, raindrops in the middle, and waves at the new bottom (near the cap).

(4) Add one drop of food coloring to the bottle.

(5) Fill the bottle about one-third with water.

(6) Screw on the cap tightly.

(7) Flip over the bottle and place it into the cup.

(8) Put it in a sunny location with ice cubes placed on top.

(9) Check the bottle after about one hour.

(10) Observe the condensation and precipitation.

THE SCIENTIFIC METHOD

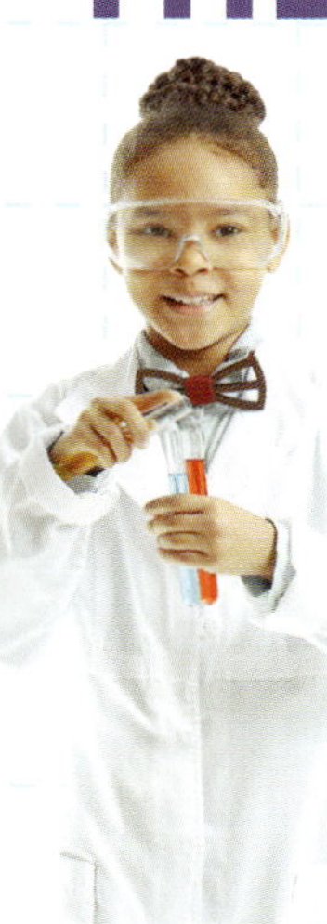

- Ask a question.
- Gather information and observe.
- Make a hypothesis or guess the answer.
- Experiment and test your hypothesis, or guess.
- Analyze your test results.
- **Modify** your hypothesis, if necessary.
- Make a conclusion.

SCIENTIST SPOTLIGHT

Heston Marc Blumenthal is a British celebrity chef, TV personality, and food writer. He likes to use science to prepare dishes by changing the matter that makes up food in unusual ways. Heston Blumenthal came to public attention with recipes such as bacon and egg ice cream and snail porridge. He is famous for popularizing "molecular gastronomy," a cooking technique that uses different temperatures to bring about tasty and surprising results.

GLOSSARY

chemical (KEM·i·kuhl): related to the interactions of substances

cycle (SIGH·kuhl): a series of events that are regularly repeated in the same order

destroyed (DIH·stroyd): eliminated from existence

evaporate (ih·VAP·uh·rayt): to turn from liquid into gas

modify (MAH·duh·figh): to change something to make it better suited for reaching a goal

molecules (MOL·uh·kyoolz): groups of atoms bonded together, representing the smallest fundamental unit of a chemical compound

physical (FIZ·i·kuhl): related to things that can be known through the five senses

temperature (TEM·pur·uh·chur): the degree or intensity of heat present in a substance or object

INDEX

AFTER READING QUESTIONS

1. What type of a change is crumpling paper?
2. What state of matter takes the shape of its container?
3. What is the Law of Conservation of Matter?

About the Author

Christina Earley lives in South Florida with her husband, son, and dog. Her favorite subject in school was science. She enjoys learning the science behind the world around her by asking questions such as how do roller coasters work. She loves mint chocolate chip ice cream and mermaids.

Written by: Christina Earley
Design by: Kathy Walsh
Editor: Kim Thompson

Photographs/Shutterstock: Cover © Carolyn Franks, ©Viktoriia Debopre: Cover, Pg 1, 3, 22, 23 © I FOOTAGE: Pg 4, 5 Strannik_fox/ AtlasStudio, Triff: Pg 10, 11 ©Lightspring: Pg 14, 15 ©Shutter_M: Pg 18, 19 Cihan Altun: Pg 6, 7 ©sandsun: Pg 8, 9 ©topseller: Pg 6 ©Nostagrams: Pg 9 ©anek.soowannaphoom: Pg 11 ©Praisaeng: Pg 12 ©kazoka: Pg 14 ©Nasky: Pg 16 © Cara-Foto: Pg 17©Michael Rosskothen: Pg 21©Brian Minkoff- London Pixels @Wiki, Pg 21, 22 ©santima.studio, Pg 21 ©Pixel-Shot

Library of Congress PCN Data
Changes in Matter / Christina Earley
Physical Science
ISBN 978-1-63897-120-7 (hard cover)
ISBN 978-1-63897-206-8 (paperback)
ISBN 978-1-63897-292-1 (EPUB)
ISBN 978-1-63897-378-2 (eBook)
Library of Congress Control Number: 2021945204

Printed in the United States of America.

Seahorse Publishing Company
www.seahorsepub.com

Published in the United States
Seahorse Publishing
PO Box 771325
Coral Springs, FL 33077